Green Pieces

Green From the Pond Up

A Collection of Cartoons

by Drew Aquilina

MARIPOSA
PRESS
Santa Fe, New Mexico

Green Pieces: Green from the Pond Up

Green Pieces: Green From the Pond Up copyright © 2011 by Drew Aquilina. All rights reserved. Printed in the United States. Protected under the Berne and Pan-American Copyright Conventions.

No part of this book may be used or reproduced or transmitted in any form or manner whatsoever or by any means, electronic or mechanical, including photocopying, recording, or by an information storage or retrieval system without written permission except in the case of brief quotations used in critical articles and/or reviews. Requests for permissions should be addressed to the publisher. Published in the United States by Mariposa Press. For information write Mariposa Press, 551 Cordova Road, #245, Santa Fe, New Mexico 87505.

http://www.greenpiecescartoons.com
http://www.greenpiecesbooks.com
mail@greenpiecestoons.com
greenpiecestoons@gmail.com

Publisher's Cataloging-In-Publication Data

Aquilina, Drew.

    Green pieces : green from the pond up : a collection of cartoons / by Drew Aquilina. -- Santa Fe, N.M. : Mariposa Press, c2011.

    p. ; cm.

    ISBN: 978-0-9666899-7-6

    1. Green pieces (Comic strip) 2. Wetlands--Caricatures and cartoons. 3. Environmental protection--Caricatures and cartoons. 4. Green movement--Caricatures and cartoons. 5. Ecology--Caricatures and cartoons. 6. Comic books, strips, etc. I. Title. II. Title: Green from the pond up.

PN6728.G74 A68 2011
741.5/973--dc22   1101

ACKNOWLEDGMENTS

WALLACE AND GROMIT © 2000 Nick Park. Character likenesses redrawn and referenced with special permission of Nick Park.

COPYRIGHT DISCLAIMER – Fair Use

X-FILES © 2000 is the product and property of Fox Publications. The incidental, negligible parody reference of two non-copyrighted characters of the work is not authorized by, associated or affiliated with or sponsored by Fox and is being cited without permission by Fox Productions. No copyright infringement is intended. Section 107 of the 1976 U.S. Copyright Act.

TRADEMARK DISCLAIMER – Nominative Fair Use

"TURTLE WAX – HARD SHELL FINISH" – Plastone Company, Inc., Chicago, Illinois. Turtle Wax, Inc. Nominative Fair Usage of this trademarked (expired) slogan is being used without permission and the publication of the trademark is not authorized or sponsored by or associated with Turtle Wax, Inc. No trademark infringement is intended.

"THE GOLDEN ARCHES" image – McDonald's Corporation, 1 McDonald's Plaza, Oak Brook, Illinois 60523. Nominative Fair Usage of this trademarked image is being used without permission and the publication of the trademark is not authorized or sponsored by or associated with McDonald's Corporation. No trademark infringement is intended.

"BATMAN" Image – DC Comics, Inc., 75 Rockefeller Plaza, New York, NY 10019. DC Comics, General Partnership by Assignment, 1700 Broadway, New York, NY 10019. Nominative Fair Usage of this trademarked image is being used without permission and the publication of the trademark is not authorize or sponsored by or associate with DC Comics, Inc. No trademark infringement is intended.

Green Pieces: Green from the Pond Up

Dedication

Unbelievable. I never thought I would get here. There are so many people to thank for their inspiration and support. I cannot do them all justice in this dedication, so I will have to thank them in person. A few special people must be acknowledged in print, however.

Green Pieces©: Green From the Pond Up would not be a reality but for my wife Lisa. She inspired me to draw when drawing wasn't a priority. Her love of cartoons is a joy. Her unending belief in me and support of my passion keeps me focused. Experiencing her beautiful smiles resulting from reading my material inspired me to start producing a daily feature again. I will always remember the day several years ago when I was moving the numerous flat files housing my original Bristol board cartoons and sketches. "You should publish these in a book because [something about being talented and funny – I stopped listening]," Lisa stated matter-of-factly. Fast forward to what now seems like an eternity and here we are. Amazingly enough, despite her exhaustive edits and revisions, cleansing of millions of the 'infamous' errant pixels and my long nights toiling at the drafting table, we are still very happily married.

I would also like to thank my family for creating an atmosphere of humor that I hope is reflected in this work. They all have been very supportive. I thank Dad for the gift of telling a great joke and Mom for the gift of understanding the 'straight line.' My oldest brother Jim taught me the fine art of the pun and dry wit and my other older brother Brett taught me the importance of humor in terms of context and content. I thank my lifelong friends Marc, Mike and Don. They played a major role in my cartooning career. Marc and I grew up together and I have learned a lot from him. Mike always asked how 'Iggy and the boys' were doing, and his interest triggered me to add drawings of my pet turtle and lizards to the bottom of my college letters. Don was especially helpful because of his ability to not understand my cartoons. I brought him material I thought was hilarious and he would rarely get the punch line. Demoralized, I would go back to the drawing board and rework the dialogue. Even though he repeatedly told me that most cartoons in general were not funny to him, it made me believe if I could draw a cartoon strip that Don would get, I would have it made. Finally, I did, he did and the rest, well, that story is still being written.

I extend a special thanks to Joseph Sardone, my extraordinary fifth grade teacher, educator and lifelong mentor. I acknowledge the University of Massachusetts Amherst and its newspaper *The Collegian* for seeing the merits of my early work, *Iggman on Campus*, and publishing it daily for several semesters. This initial training helped me understand that successful cartooning involves deadlines. I also acknowledge *The Echo*, Western Connecticut State University's student newspaper and *The Newtown Bee* in Newtown, Connecticut. It was while contributing to *The Newtown Bee* that I partnered with my friend Deb aka Dagny, a very patient and gifted writer. She taught me to appreciate the art and craft of writing humor. I also extend a special thanks to *The News-Times* in Danbury, Connecticut and its late Editor, Edward Frede. Mr. Frede's kindness and confidence in me changed my life and I will be forever in his debt.

Finally, I want to thank my mentors, Brian Walker and the late Gill Fox, as well as the rest of my friends at the Monthly Lunch in Bethel, Connecticut for their friendship and guidance. Because of my exposure in *The News-Times*, I received a call from Gill Fox who wanted to know "who the heck" I was (being a relative unknown cartoonist in a very exclusive profession). Mr. Fox told me that a few cartoonists met monthly at a local restaurant in Bethel and he invited me to join the group for lunch. I couldn't believe that Gill Fox asked me to meet for lunch and needless to say I was very nervous. I attended the luncheon and was able to meet cartoonists whose work I grew up with. They were very gracious to me, this young guy struggling with his deadlines. It was here that I was able to talk (and mostly listen) to these professionals, the Who's Who from the cartoon page, *The New Yorker* magazine and independent illustrators and writers. I surprisingly learned I was dealing with similar cartooning issues and challenges. These meetings and exchanges set me on the course to challenge my cartooning creativity and further develop the true "Nature" of *Green pieces*.

Green Pieces: Green from the Pond Up

Green Pieces: Green from the Pond Up

FOREWORD
by Brian Walker

I started working at the Museum of Cartoon Art in 1974, a few months after I graduated from college. On numerous occasions, aspiring cartoonists would bring their work to the Museum, particularly on Sunday afternoons when my father, Mort Walker, was manning the gift counter. I was amazed at how patient Mort was with these young artists. He explained to me that when he was growing up, his father would often take him to the offices of the *Kansas City Star*. The guys in the art department took the eager kid under their wing and gave him tips on how to improve his drawing. My father was convinced that this encouragement helped him to become a successful cartoonist and felt an obligation to help other young artists in a similar way. This story left a strong impression on me.

Beginning in the mid-1970s, we offered a cartoon course at the Museum and some of our former students went on to pursue successful careers. In the early 1980s, I agreed to give private lessons to a finalist in one of our cartoon contests, Mark Tonra. I took him to meet some of his idols, including Dik Browne, and helped him to produce a short animated film. Mark became a professional magazine cartoonist and eventually sold three different comic strips to three different syndicates. I continued working with young cartoonists in the 1990s, teaching courses in humor writing at Fairfield University in Connecticut and the School of Visual Arts in New York City. This was when Gill Fox introduced me to Drew Aquilina.

At that time, Drew's cartoons were being published in the *Danbury News-Times*. Drew visited my class at Fairfield U. and Gill and I encouraged him to continue his career. I always told my students that the most important goal when starting out was to get their drawings published, even if they had to work for free. This is how cartoonists learn to communicate with their readers. In the years since I met him, Drew has continued to draw for publication and build his audience.

Drew's technique is now fully developed. His writing is well-paced and conveys a positive, environmental message. The backgrounds depict a realistic locale and the scenery documents the changing seasons. The main cast, Iggy, Radic, Cabby, and Roc, are charming and believable. The humor is warm and genuine. The strips in this book clearly demonstrate Drew's talents as an artist.

Selling a strip to one of the major syndicates is more difficult than it has ever been, but I think Drew has a good chance to achieve national recognition. In the meantime, he has been honing his craft, waiting for his big break. This book is a major step in that direction. I am flattered that he has asked me to write the foreword and I am pleased that I was able to contribute to his progress. As my father taught me many years ago, helping aspiring cartoonists is the least we can do to return the good fortune we have enjoyed working in this wonderful business. I wish Drew success with *Green Pieces* and I hope he will help other artists who follow in his footsteps.

Green Pieces: Green from the Pond Up

Green Pieces: Green from the Pond Up

Green Pieces: Green From the Pond Up

Welcome to the Pond

Winner, 2010 Green Book Festival, San Francisco, CA, Honorable Mention, Comics/Graphic Novels

"All in all this is a wonderful book, full of top-notch illustrations and adorable characters that will bring nature and the inhabitants alive to you in a superb way. Very well done and highly recommended."
~ Shirley Johnson, Senior Reviewer, *Midwest Book Review*

"To create a work that is a hit with both youngsters and adults is a very difficult task. . . . A true master of the craft can produce a work that appeals to young and old alike - Wind In The Willows and Alice In Wonderland are two examples. . . . I see Green Pieces as being such a book. . . . Clearly this man is a creative and very humorous genius."
~ Simon Barrett, *Blogger News Network*

"This is by far the most entertaining and instructive cartoon book I've ever encountered and, as such, is a 'must have' for children of all ages, school libraries and 'young at heart' adults."
~ Dr. Jana S. Eaton, Educator, Artist and Author of numerous academic publications, Green Valley, AZ

"Aquilina's whimsical and clever cartoon strips are entertaining and thoughtful with creative illustrations that will help bring you out of your shell with a smile."
~ Joseph Birkett, Publisher/Editor, *Tubac Villager*

"Drew Aquilina was going green before going green was cool. If you've ever wondered what a naked turtle looks like, *Green Pieces* is the cartoon series for you."
~ Joe Devito, Famous Comedian, www.joedevito.com

Iggy

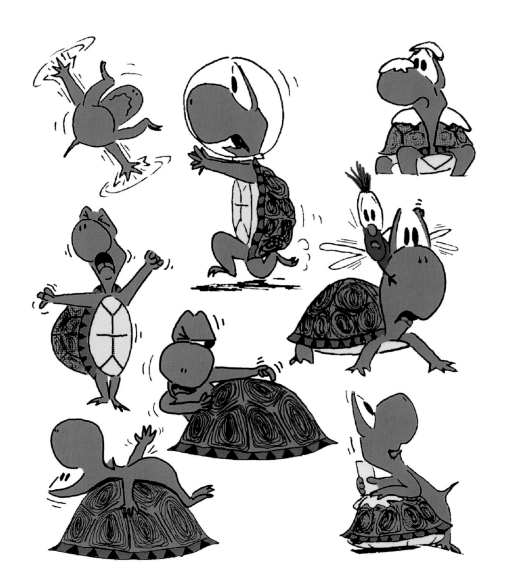

Cleverly yet cautiously approaching the world around him, Iggman the turtle seeks the simple pleasures of life. Iggy (as his friends call him) has a unique perspective on his environment. As a turtle, his species' longevity promotes wisdom. It is Iggy's naiveté, however, that gives away his true age. His shell, the seemingly magical carapace that follows Iggy everywhere, is the center of attention. The shell serves as Iggy's sanctuary as well as his burden. Unusual for turtles, Iggy battles bouts of claustrophobia. He tempers his affliction by frequently giving the outside of his shell a good coat of wax.

"This collection of cartoons is a must for young and old. . . . I love to imagine myself as part of the Pond friends and would enjoy their many antics first hand. I am confident parents and children alike will agree."
~ Theresa Eberhard Asch, Educator, Danbury, CT

"Wit and whimsy can lead to wisdom or at least awareness. In *Green Pieces*, Drew Aquilina skillfully uses cartoon humor to parody human foibles and tweak consciousness of nature and our environment."
~ Marge Hanley, Retired Writer/Educator, *The Indianapolis News*

"Teaching my special needs students an awareness of ecology and an attentiveness to the environment is always a challenge. The characters in this book are relatable to both young and old and the situations are springboards to discussion. Even my non-readers want to look and learn from the adventures of Iggy, Radic, Cabby and Roc."
~ Elyse Blum, Exceptional Education Teacher, Hialeah High School, Hialeah, Florida

"These environmental cartoons are perfect to teach school children about respecting the environment."
~ Laura Chamberlain, *Tampa Bookworm Blog*

Radic

Radic the dragonfly epitomizes the word "bug." Making his rounds at the pond and through nature, Radic stirs up trouble wherever he flies. This pesky but lovable cad is the self-proclaimed "best friend" of Iggman the turtle. It is unknown whether Radic truly seeks Iggy's friendship or simply covets Iggy's warm, comfy-cozy shell on which to land and sleep. Regardless, together they embark on mischievous adventures while building an endearing "co-existence."

"Green From the Pond Up creates a 'humanized' image of nature's creatures. His cartoons give insightful, yet colorful, characterizations as vivid as those pictured on film. BRAVO! You 'Drew' me into your world!"
~ William D. Mast, Renowned Travel Photographer, Mesa, AZ

"It's a fun way of increasing awareness about the environment, of teaching kids, and adults, to respect nature, seeing the world from the viewpoint of these quirky characters."
~ Carol Evans, *Carol's Notebook Blog*

"I found this collection of cartoons quite amusing. ... Some of the strips had me laughing out loud, hilarious. ... quite cute and several had a 'current events' theme to them (like the spotted owl and its loss of habitat). This was a great way to pass the time, curled up in a chair with a cup of coffee."
~ Diane Hoffmaster, *Turning The Clock Back Blog*

"This book is a great way to educate the public, especially children, about the importance of protecting nature rather than harming it. ... I would definitely recommend parents buy this book for their children and teachers to buy it for their classrooms."
~ *The Book Worm Blog*

Cabby

Cabby the techno-bullfrog had his DNA altered by pollution during the critical tadpole stage. As a result, Cabby's brain developed instead of his legs. Cabby's genius lies in the scientific realm with heavy emphasis on computers and physics. While normally focused on his scientific pursuits, from time to time Cabby will indulge in less lofty aspirations by helping his friends get in and out of trouble.

"The cartoon's simplicity is its winning factor. Its humour will be enjoyed by young adults and their parents. Kids will enjoy it for the drawings but the parents will enjoy the humour as representative of our times in this banal world. I have to endorse the bats most."
~ PK Reeves, *Aisle B*, Canada

"This book had two things I really enjoyed, comics and an educational look at our environment without letting people know they are being educated."
~ Sandra Stiles, *Musings of A Book Addict Blog*

"Green Pieces reminds us, often with hilarious results, that we share this world, and would be wise to keep it Green from the Pond Up."
~ Becky Reyes, Singer/songwriter, Rio Rico, AZ

"Very fresh and fun with characters you'll just have to love. A must read for all ages."
RATING: ♥♥♥♥♥
~ *A Moment with Mystee Blog*

"Drew can draw. He has also accomplished the unlikely and made ecological commentary likeable. His collection of pond parables touch lightly and yet still makes waves. Dive into *Green Pieces: Green From the Pond Up* and you will come out smiling."
~ Gary Isaacson, Author, *Game Over* (2007) and *Crude Customs* (July 2010)

Roc

Roc is the friendly neighborhood raccoon. Charming and endearing to a fault, Roc possesses an uncanny ability to recognize the reality of most situations. As the local rogue, he gets through life by living out of dumpsters and trash cans and picking up odd jobs from Cabby. Roc reminds you of the uncle that families don't talk about much. When he is not sleeping off a night of heavy rubbish rummaging, Roc can be a very good friend and fun companion.

Spring

Green Pieces: Green from the Pond Up

Green Pieces: Green from the Pond Up

Green Pieces: Green from the Pond Up

 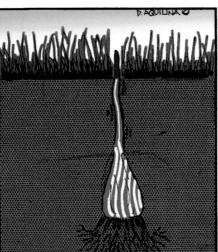  

JUST REMEMBER, EVERY FLOWER THAT EVER BLOOMED HAD TO GO THROUGH A WHOLE LOT OF DIRT TO GET THERE!

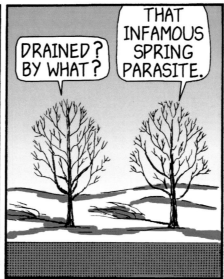

Green Pieces: Green from the Pond Up

Green Pieces: Green from the Pond Up

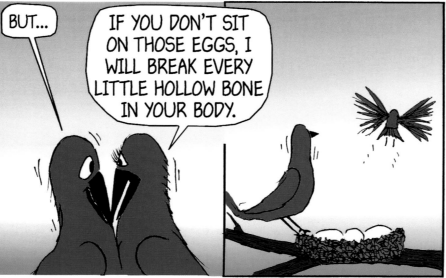

Green Pieces: Green from the Pond Up

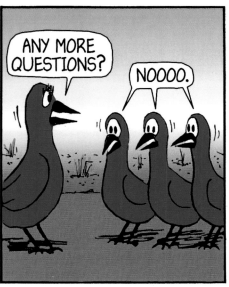

Green Pieces: Green from the Pond Up

Green Pieces: Green from the Pond Up

Green Pieces: Green from the Pond Up

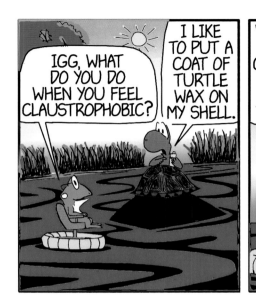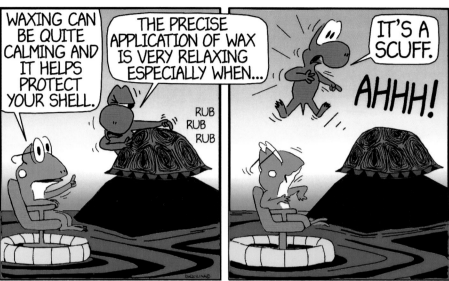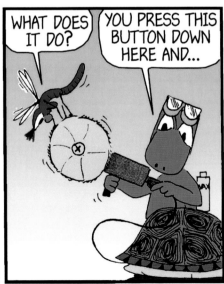

Green Pieces: Green from the Pond Up

Green Pieces: Green from the Pond Up

Green Pieces: Green from the Pond Up

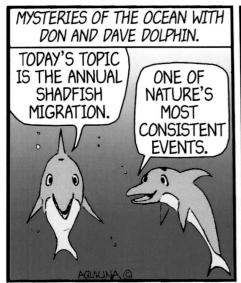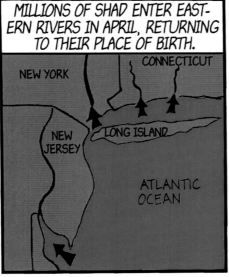

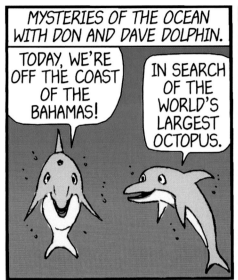

Summer

Green Pieces: Green from the Pond Up

Green Pieces: Green from the Pond Up

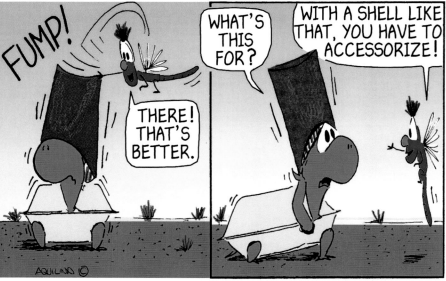

Green Pieces: Green from the Pond Up

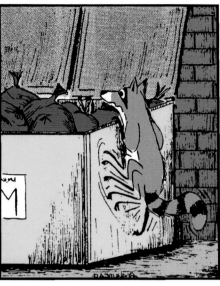

Green Pieces: Green from the Pond Up

Green Pieces: Green from the Pond Up

Green Pieces: Green from the Pond Up

Green Pieces: Green from the Pond Up

Green Pieces: Green from the Pond Up

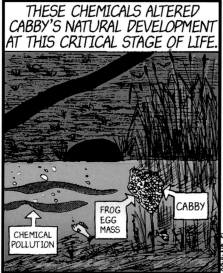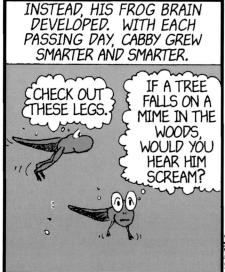

Green Pieces: Green from the Pond Up

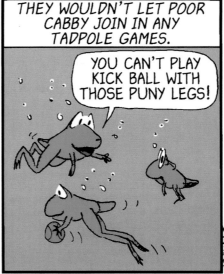

Green Pieces: Green from the Pond Up

Green Pieces: Green from the Pond Up

Green Pieces: Green from the Pond Up

Green Pieces: Green from the Pond Up

Green Pieces: Green from the Pond Up

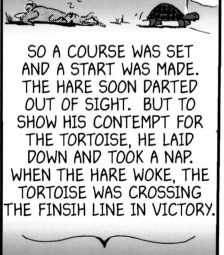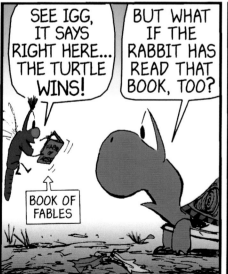

Green Pieces: Green from the Pond Up

Green Pieces: Green from the Pond Up

Green Pieces: Green from the Pond Up

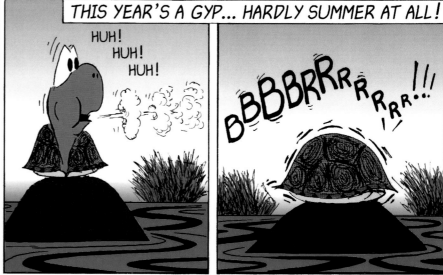

Green Pieces: Green from the Pond Up

Green Pieces: Green from the Pond Up

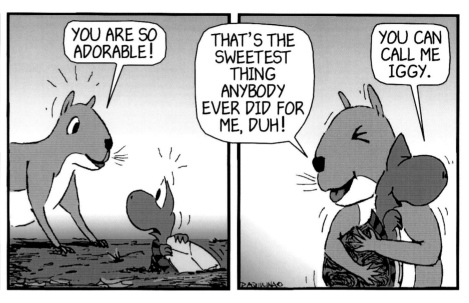

Green Pieces: Green from the Pond Up

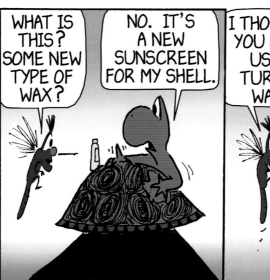

Green Pieces: Green from the Pond Up

IN TERMS OF 'QUALITY OF LIFE,' A TURTLE'S EXISTENCE MUST RANK HIGH AMONG LEISURE-SEEKING LIFE FORMS. WITH VERY FEW ENEMIES, IGG ...

DEVOTES MOST OF HIS DAY TO EATING, SUN BATHING AND THE OCCASIONAL SWIM.

IGG, WHO'S GOT IT BETTER THAN YOU?

NOBODY!

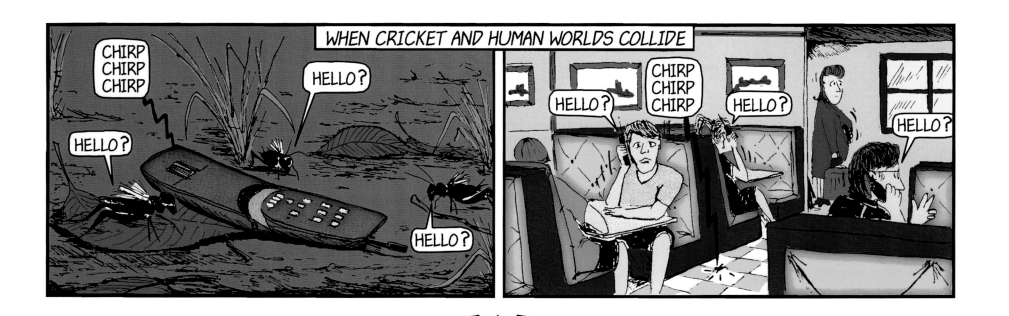

Green Pieces: Green from the Pond Up

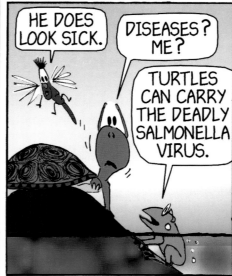

Green Pieces: Green from the Pond Up

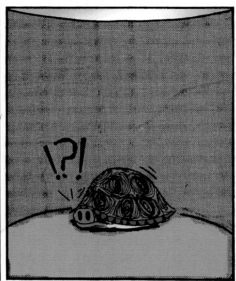

Green Pieces: Green from the Pond Up

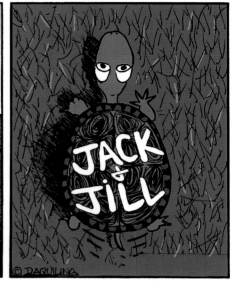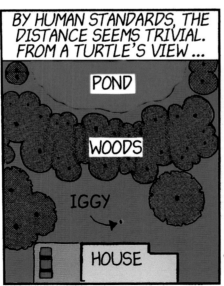

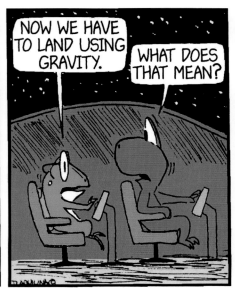

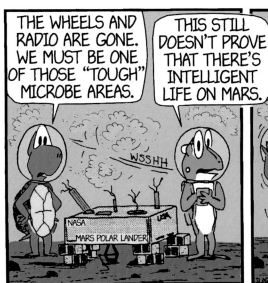

Green Pieces: Green from the Pond Up

Fall

Green Pieces: Green from the Pond Up

Green Pieces: Green from the Pond Up

Green Pieces: Green from the Pond Up

Green Pieces: Green from the Pond Up

Green Pieces: Green from the Pond Up

Green Pieces: Green from the Pond Up

Green Pieces: Green from the Pond Up

Green Pieces: Green from the Pond Up

Green Pieces: Green from the Pond Up

Green Pieces: Green from the Pond Up

Green Pieces: Green from the Pond Up

Green Pieces: Green from the Pond Up

Green Pieces: Green from the Pond Up

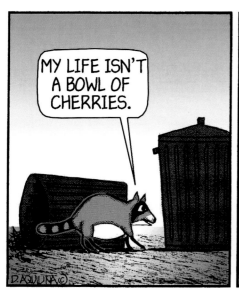

Green Pieces: Green from the Pond Up

Green Pieces: Green from the Pond Up

Green Pieces: Green from the Pond Up

Green Pieces: Green from the Pond Up

Green Pieces: Green from the Pond Up

165

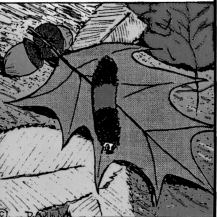

Winter

Green Pieces: Green from the Pond Up

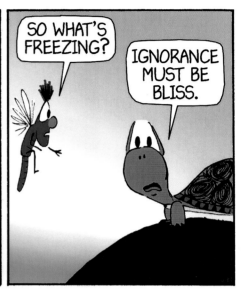

Green Pieces: Green from the Pond Up

Green Pieces: Green from the Pond Up

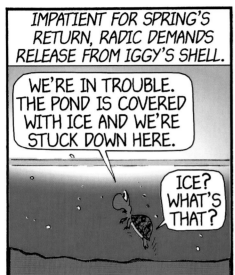

Green Pieces: Green from the Pond Up

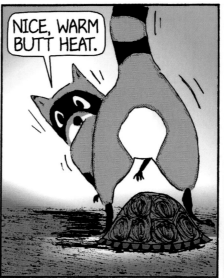

Green Pieces: Green from the Pond Up

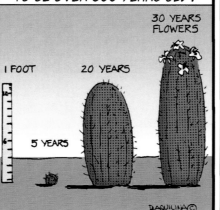

Green Pieces: Green from the Pond Up

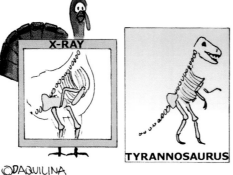

Green Pieces: Green from the Pond Up

Green Pieces: Green from the Pond Up

Green Pieces: Green from the Pond Up

Green Pieces: Green from the Pond Up

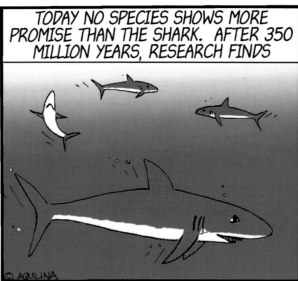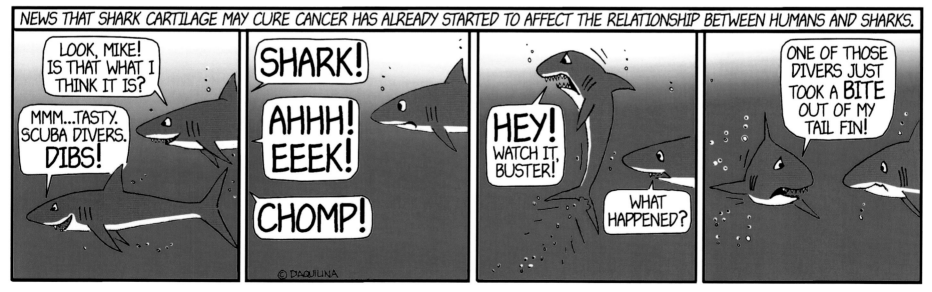

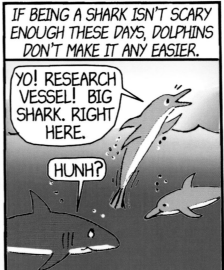

Green Pieces: Green from the Pond Up

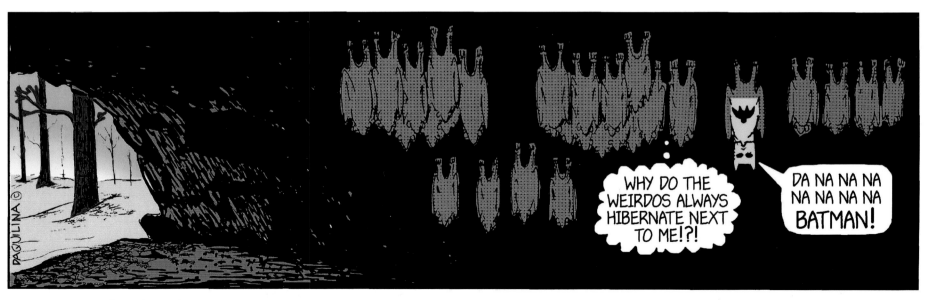

Green Pieces: Green from the Pond Up

THE POND ORDER FORM

OR ORDER AT WWW.GREENPIECESCARTOONS.COM

ITEM	QTY	UNIT PRICE	TOTAL
GREEN PIECES: GREEN FROM THE POND UP (ORDER 10 OR MORE BOOKS AND RECEIVE 15% DISCOUNT) (DISCOUNT IS NOT APPLIED TO SHIPPING)		$19.99	$
		SUBTOTAL	$
* 9.00% SALES TAX - ON ALL ORDERS ORIGINATING IN ARIZONA.		*TAX	$
**$8.00 OR 10% OF THE TOTAL - WHICHEVER IS GREATER. GROUND SHIPPING. ALLOW 5-7 DAYS FOR DELIVERY.		**SHIPPING	$
MAIL ORDER FORM TO: GREEN PIECES, 7904 E. CHAPARRAL ROAD, #A110-496, SCOTTSDALE, AZ 85250-7210		TOTAL	$

NAME:

ADDRESS:

CITY, STATE, ZIP:

DAYTIME PHONE:

EMAIL:

METHOD OF PAYMENT (CIRCLE ONE):

VISA   MASTER CARD   DISCOVER   AMERICAN EXPRESS

ACCOUNT NUMBER   EXPIRATION DATE

SIGNATURE   3-4 DIGIT SECURITY NUMBER